故園畫憶

庚寅中秋
韩启德题

故园画忆系列
Memory of the Old Home in Sketches

上海老洋房

Old Foreign-style Buildings in Shanghai

胡家康　绘画　底谓　胡家康　撰文

Sketches by Hu Jiakang
Notes by Di Wei & Hu Jiakang

學苑出版社
Academy Press

图书在版编目（CIP）数据

上海老洋房 / 胡家康绘画；底谓、胡家康撰文. — 北京：学苑出版社, 2015.9

（故园画忆系列）

ISBN 978-7-5077-4866-6

Ⅰ. ①上… Ⅱ. ①胡… ②底… Ⅲ. ①建筑画—钢笔画—作品集—中国—现代

Ⅳ. ①TU-881.2

中国版本图书馆CIP数据核字(2015)第212884号

出 版 人：孟　白
责任编辑：沈　萌　周　鼎
出版发行：学苑出版社
社　　址：北京市丰台区南方庄2号院1号楼
邮政编码：100079
网　　址：www.book001.com
电子信箱：xueyuanpress@163.com
销售电话：010-67601101（销售部）、67603091（总编室）
经　　销：全国新华书店
印 刷 厂：河北赛文印刷有限公司
开本尺寸：889 × 1194　1/24
印　　张：7.75
字　　数：15千字
图　　幅：158幅
版　　次：2015年11月北京第1版
印　　次：2020年8月北京第2次印刷
定　　价：45.00元

目　录

自　序　　胡家康

安福路、长乐路、多伦路等

吴国桢住宅　3
豫园　4
安亭路西班牙式花园住宅　5
玉佛禅寺　6
天主教望德堂　7
爱文公寓　8
周宗良住宅　9
张爱玲居所　10
蒲园　11
潘宗周住宅　12
长乐路古典式花园住宅　13
沪西礼拜堂　14
摩西会堂　15
永丰村点式公寓　16
吕班公寓　17
邹韬奋故居　18
圣伯多禄堂　19
大境关帝庙　20
杜公馆　21
圣沙忽略堂　22
爱庐　23
鸿德堂　24
李观生住宅　25
孔祥熙住宅　26

汾阳路、复兴中路、华山路等

木结构独立式花园洋房　29
犹太人俱乐部　30
丁贵堂住宅　31
小白宫　32
白公馆　33
麦琪公寓　34
复兴西路英国乡村式住宅　35
复兴西路花园住宅　36
诸圣堂　37
柳亚子旧居　38
克莱门公寓　39
阿麦伦公寓　40
荣德生旧居　41
张学良公馆　42
圣尼古拉教堂　43
沪东礼拜堂　44
国际礼拜堂　45
虹桥路英式花园别墅　46
沙逊别墅　47
贺子珍旧居　48
熊佛西楼　49

孙家花园 50
丁香花园 51
郭棣活住宅 52
丁香别墅 53
嘉色喇住宅 54

华亭路、淮海中路、巨鹿路等

华亭路地中海式花园住宅 57
华亭路英式花园住宅 58
佛兰克林住宅 59
苏联驻沪商务代办处 60
席宅 61
赵丹故居 62
上方花园 63
巴塞住宅 64
盛宣怀住宅 65
逸邨 66
何应钦住宅 67
淮海中路花园里弄住宅 68
诺曼底公寓 69
宋庆龄故居 70
贝家老宅 71
甘村新式里弄住宅 72
法国太子公寓 73
懿园 74
王时新住宅 75
圣三一基督教堂 76
刘吉生故居 77
巨鹿路英国乡村式花园住宅 78
巨鹿路英式双毗连花园住宅 79
康平路花园住宅 80

瑞金二路、陕西北路、铜仁路等

焉息堂 83
景灵堂 84
真如寺 85
新天地石库门 86
龙华古寺 87
清心堂 88
新天安堂 89
徐家汇天主堂 90
周湘云住宅 91
吴妙生住宅 92
四明公所 93
英商马立斯住宅 94
三井洋行大班住宅 95
托益住宅 96
怀恩堂 97
西摩会堂 98
步高里 99
若瑟堂 100
袁佐良寓所 101
思南公馆别墅群 102
贺绿汀旧居 103
陈家巷乡村 104
卫乐园 105
马歇尔公寓 106
露德圣母堂 107
张叔驯住宅 108
史量才旧居 109
铜仁路毗连式公寓 110
吴同文花园住宅 111
邱氏住宅 112

武康路、新华路、兴国路等
正广和洋行大班住宅 115
巴金故居 116
原意大利总领事官邸 117
黄兴故居 118
朱敏堂住宅 119
东方汇理银行大班故居 120
武夷路花园别墅 121
沐恩堂 122
龚品梅故居 123
孙中山故居 124
西城回教堂 125
新华路德式居民别墅 126
李佳白住宅 127
新华路英式乡村别墅 128
新华路花园住宅 129
新华路西班牙式花园住宅 130
周均时住宅 131
梅泉别墅 132
东正教圣母大堂 133
英商太古洋行大班住宅 134
兴国路英国维多利亚滨海建筑 135
兴国宾馆 6 号楼 136

延安西路、永嘉路、愚园路等
孙科住宅 139
延安西路西班牙式花园住宅 140
延安中路英侨住宅 141
马勒别墅 142
永业大楼 143
延庆路法国古典式花园住宅 144
布哈德住宅 145
永福路西班牙式花园住宅 146
孔祥熙旧宅 147
荣智勋住宅 148
永嘉路花园住宅 149
宋子文旧居 150
外国弄堂“雷米坊” 151
岳阳路现代式花园住宅 152
宋子文旧宅 153
岳阳路现代式花园别墅 154
霖生医院旧址 155
陈楚湘住宅 156
蒋光鼐旧居 157
王伯群住宅 158
董竹君住宅 159
新华村 160
西本愿寺 161

闵行区、青浦区、崇明县等
法华塔 165
七宝古镇 166
南张天主堂 167
朱家角古镇 168
泰来桥天主堂 169
练塘灵恩堂 170
蔡家湾天主堂 171
佘山天主教堂 172
黄家花园 173

Contents

Preface Hu jiakang

Anfu Road, Changle Road, Duolun Road, etc.

Former Residence of Wu Guozhen 3
Yu Garden 4
Spanish Garden House on Anting Road 5
Yufo Temple 6
The Augustinian Procuration of the Catholic Church 7
Avenue Apartment 8
Former Residence of Zhou Zongliang 9
Former Residence of Zhang Ailing 10
Garden of Bourgeat 11
Former Residence of Pan Zongzhou 12
Classic Garden House on Changle Road 13
West Shanghai Church 14
Ohel Moishe Synagogue 15
An apartment in Yongfeng Village 16
Lvban Apartment 17
Former Residence of Zou Taofen 18
Saint Peter's Hall 19
Guan Yu Temple on Dajing Road 20
Former Residence of Du Yuesheng Mansion 21
Saint Xavier Church 22
Ai Lu 23
Fitch Memorial Church 24
Former Residence of Li Guansheng 25
Former Residence of Kung Hsiang-hsi 26

Fenyang Road, Middle Fuxing Road, Huashan Road, etc.

Free-Standing Garden Villa with a Wooden Structure 29
Jewish Club 30
Former Residence of Ding Guitang 31
The Little White House 32
The Bai Mansion 33
Magy Apartment 34
English Country-Style Residence on West Fuxing Road 35
Garden House on West Fuxing Road 36
All Saint's Church 37
Former Residence of Liu Yazi 38
Kremen Apartment 39
Amyron Apartments 40
Former Residence of Rong Desheng 41
Mansion of Zhang Xueliang 42
Saint Nicholas Church 43
East Shanghai Church 44
Shanghai Community Church 45
British-Style Garden House on Hongqiao Road 46
Sassoon Villa 47
Former Residence of He Zizhen 48
Xiong Foxi Building 49
Sun Garden 50
The Lilac Garden 51
Former Residence of Guo Dihuo 52
Lilac Villa 53
Former Residence of Leopold Cassella 54

Huating Road, Middle Huaihai Road, Julu Road, etc.

Mediterranean Garden House on Huating Road 57
British Garden House on Huating Road 58
The Former Franklin Residence 59
Former Commercial Agency of the Soviet Union in Shanghai 60
Former Residence of Xi Zhengfu 61
Former Residence of Zhao Dan 62
Shangfang Garden 63
The Former Basset Residence 64
Former Residence of Sheng Xuanhuai 65
Yi Cun 66
Former Residence of He Yingqin 67
Lane-and-Alley Garden Houses on Middle Huaihai Road 68
Normandy Apartment 69
Former Residence of Soong Ching-ling 70
The Old Bei Family House 71
Newly-Developing Lanes and Alleys, Gancun Village 72
Former Residence of a French Prince 73
Yi Garden 74
Former Residence of Wang Shixin 75
Holy Trinity Cathedral 76
Former Residence of Liu Jisheng 77
Garden House on Julu Road 78
British-Style Double Garden House on Julu Road 79
Garden House on Kangping Road 80

2nd Ruijin Road, North Shaanxi Road, Tongren Road, etc.

Catholic Country Church 83
Jingling Church 84
Zhenru Temple 85
Stone Houses near Shanghai New World 86
Longhua Ancient Temple 87
Pure Heart Church 88
The Union Church 89
Xujiahui Catholic Church 90
Former Residence of Zhou Xiangyun 91
Former Residence of Wu Miaosheng 92
Siming Association 93
Former Residence of Maris 94
Residence for the Senior Group of the Mitsui Bussan Kaisha Co., Ltd 95
Former Residence of Toy 96
Grace Baptist Church 97
Ohel Rachel Synagogue 98
Cité Bourgogne 99
St. Josephis Church 100
Property of Legendary Yuan Zuoliang 101
Sinan Mansion 102
Former Residence of He Lvting 103
Chenjiaxiang Village 104
Willow Garden 105
Marshall Mansion 106
Our Lady of Lourdes Church 107
Former Residence of Zhang Shuxun 108
Former Residence of Shi Liangcai 109
Adjoining Apartments on Tongren Road 110
Garden House of Wu Tongwen 111
Qiu Mansion 112

Wukang Road, Xinhua Road, Xingguo Road, etc.

Former Residence for the Senior Group of

Aquarium Bank 115
Residence of Ba Jin 116
Former Italian Consulate General 117
Former Residence of Huang Xing 118
Former Residence of Zhu Mintang 119
Former Residence of the Senior Group of the Banque de L'indochine 120
Garden Villa on Wuyi Road 121
Moore Memorial Church 122
Former Residence of Gong Pinmei 123
Former Residence of Sun Yatsen 124
West Shanghai Mosque 125
German-Style Villa on Xinhua Road 126
Former Residence Of Li Jiabai 127
British Country Style Villa on Xinhua Road 128
Garden House on Xinhua Road 129
Spanish Garden House on Xinhua Road 130
Former Residence of Zhou Junshi 131
Meiquan Villas 132
Russian Orthodox Church 133
Residence for the Senior Group of the Butterfield & Swire Company 134
British Victorian Buildings in the Coastal Area on Xingguo Road 135
Building No.6, Radisson Plaza Xing Guo Hotel Shanghai 136

West Yan'an Road, Yongjia Road, Yuyuan Road, etc.

Former Residence of Sun Ko 139
Spanish Garden House on West Yan'an Road 140
Former Residence of a British-Born Chinese gentleman on Middle Yan'an Road 141
Moller Villa 142
Yongye Building 143
Classic French Garden House on Yanqing Road 144
Former Residence of Buchard 145
Spanish Garden House on Yongfu Road 146
Former Residence of Kung Hsianghsi 147
Former Residence of Rong Zhixun 148
Garden House on Yongjia Road 149
Former Residence of Soong Tse-ven 150
Remi Apartments, Foreign Lane 151
Modern House on Yueyang Road 152
Former Residence of Soong Tse-ven 153
Modern Garden on Yueyang Road 154
Former Linsheng Hospital 155
Former Residence of Chen Chuxiang 156
Former Residence of Jiang Guangnai 157
Former Residence of Wang Boqun 158
Former Residence of Dong Zhujun 159
Xinhua Village 160
West Honganji Monastery 161

Minghang District, Qingpu District, Chongming County, etc.

Fahua Tower 165
Qibao Anccient Town 166
Nanzhang Cathedral 167
Zhujiajiao Ancient Town 168
The Catholic Church by Tailai Bridge 169
Ling'en Church in Liantang 170
CAI Home Bay Church 171
Sheshan Cathedral 172
Garden of the Huang Family 173

自　序

上海——一座走在中国最前沿的城市，无论过去还是今天，在世界舞台上时时展现着它独特的风采，吸引着海内外有识之士涌向这片乐土，追寻适宜的高品质生活。

踏入大上海，一条黄浦江形成了隔江相望的两岸——浦东和浦西。一边是陆家嘴金融贸易区不断刷新高度的现代建筑，一边是外滩老城区有厚重历史感的万国建筑群，今日豪华与昔日繁荣遥相呼应，它们虽处同城，但发展的轨迹却是那样的不同。

从浦东江边回望浦西，那是散发着另一种气息的上海。浦西，上海本帮文化的发源地。大上海，小里弄，这座城市就好像一个有生命的肌体。其自开埠以来，就如此令人向往。旧梦萦绕，那历史的沧桑历历在目：建于20世纪初，风格多样、造型优美的老洋房已经成了上海的一个符号。岁月的延伸，让我去探秘上海滩老洋房的前世今生，体验上海经典弄堂的富贵及颓败。

花园洋房是上海最经典的住宅，每一幢都承载着一个海上当年的传奇故事，它可以被视作为一座成功人士的纪念碑，见证着近代上海的变迁。那一条条沧桑的老街道，那一栋栋陈旧的老洋房，那一排排凹凸有致的石库门，那一家家情浓意密老酒吧，仍然能让我们穿越时空隧道，去怀念老上海曾经的繁华。旧上海的十里洋场，造就了众多风格各异的欧美建筑：砖瓦堆砌的洋房透出特有的质朴气息，落地的窗户散发出西式的氛围，庭院的绿树让洋房在郁郁葱葱中摇曳着风姿。这些不同文化历史背景的老洋房作为上海最经典的住宅形式，已然成了上海人心中一个挥之不去的情结。

20世纪二三十年代的上海老洋房，其设计的理念是：以大树、高墙和灌木构筑起街区条块；用湖泊、河流等水系与自然植栽的串连，形成乡村意象；将稳重和优雅的欧洲风格建筑符号混搭在一起，于风格和谐之中彰显多姿的个性；极富装饰意味的地铺、泳池、铸铁栏杆等建筑小品，凸显了完美的老洋房印象。老建筑是一座城市的物质记忆，承载着太多的历史文化。许多分布在上海各处的老洋房、老别墅，充满了20世纪早期各种风格建筑精品的语汇。其住宅类型有：石库门、高层公寓、独立式花园住宅、联排式住宅、联立式花园住宅、新式里弄住宅、现代公寓式住宅、外廊式住宅、带内院独立式花园住宅等十多种，堪称“历史建筑博物馆”。

我生在上海，居住在“步高里”石库门旁的弄堂里，所借画室即为建国路上我大姐的旧居。此画室也是老洋房建筑的衍生物——新式里弄房。我小时候与周边邻居家的小孩一起上学、读书、玩耍；长大后生活工作仍在这座城市，故对上海旧城区内各处较强的区域性地标建筑相当熟悉，对浦西这片土地怀有深深的眷恋。著名的新天地和思南公馆就在我从小生活的原卢湾区；衡山路酒吧一条街在邻近的徐汇区；新华路、虹桥路上风姿卓越的老洋房在长宁区；铜仁路上的西洋建筑在静安区；还有那座落在多伦路、山阴路上的名人名宅则集中在虹口区。

旧时的上海除闸北和原南市两区外都是租界地，洋泾浜（爱多业路，即今日延安路）以北是英租界，以南是法租界，虹口一带是日租界。租界管理制度下的上海，呈现了不同风格的万国建筑。这些不同文化历史背景的老洋房建筑，铸就了上海的繁华兴盛、摩登时尚的独特气质，比起一些风平浪静地走过千百年历史的城市，拥有这段特殊历史的上海，显得更加风情万种。

红瓦屋顶，赭色百叶窗，小巧的阁楼，镂空雕花的欧式铁门，饱经风雨的黑色铁栅栏，营造了整体宁静、优雅的氛围。幽深的后花园，几棵参天大树从墙头探出，隐约可见精美的屋顶或阳台。花园深处时不时会传出欢快的爵士乐。而当您不经意地踩响砖道落叶，或许还会听到一股流水般的钢琴或小提琴的浪漫旋律，亦或是用老式唱机之大喇叭播放的若有若无、耐人寻味的老歌，即如周璇的《永远的微笑》、崔萍的《今宵多尊重》、潘秀琼的《情人的眼泪》、葛兰的《卡门》乃至《夜上海》、《茉莉花》、《我要你的爱》等不一而足。在浓郁的梧桐树后，那一座座幽静的花园里，幢幢风格迥异、各呈奇姿的小洋楼若隐若现。这就是上海的老洋房。

旧上海的风貌及男男女女颇为西化，很多有识之士于“出国潮”后留洋而归。走在街上你会看见，在叮铛作响的有轨电车驶过后，身着旗袍的女士挽着上海“老克勒”的手臂、牵着小狗在轻盈地散步。弄堂旁停着美国制造的“克莱斯勒”、“福特”或“道奇”。车上会走出一位上海“小开”拿着手提包匆匆离去。海派文化是植根于外来文明和中国传统文明之间，在精英文化和通俗文化之间呈现出开放的姿态，吸纳消化那些外国的，主要是西方的文化元素，创立了新的富有自己独特个性的文化。其特点乃海纳百川，善于扬弃，追求卓越，锐意创新。

我对如此妩媚妖娆的老上海一直情有独钟。那历史的沧桑和风情给了我极大的冲击。我想要用敏锐的艺术眼光捕捉、用笔记录这特有的上海形象及海派文化——20 世纪早期外国人在租界地建造的西洋风格的老洋房。那些解放前属于个人所有的花园洋房，如今多为国有财产。如今一幢老洋房里往往能见到十多户人居住。而有些名人故居，已是政府保护的历史文物建筑。即使居住的后代也不能进行交易买卖。缘于历史原因，大部分老洋房都经历了革命的洗礼和岁月的冲刷。幸运的是，在多方努力之下，一些濒临毁灭的人文故居得以保存。我原计划借鉴运用大量的纪实资料，以充分再现老洋房的风采。然而，时下有些建筑已是政府或军事部门，寻常人不得入内；也有的已经改造为私人会所或住宅，探访亦同样困难，于是，我只能凭借写生和摄影多种手段，甚至通过目测、记忆乃至图片和文字记载，尽可能真实完整地还原对象。不管是刮风下雨，还是严寒酷暑，我始终坚持亲历而为，亲身感受，亲眼目睹，在取得第一手资料或直接写生时，忠实于建筑物的原貌。但仅仅这些还不够，作为一个艺术创作者，再现对象时不是照相机式的记录，还应有取有舍，既能还老宅以原貌，又能传递艺术的魅力。此画册的每一幅图都倾注了我的精心经营，其中没有半点浮夸。有些花园洋房内住户较多，建筑功能已被彻底改变，原先的停车库变为住房且砌墙开窗，阳台、窟形门窗、阳光走道封闭，违章搭建比比皆是，我只能拚弃照相式的写真记录，不得不像导演一样，以自己的理解加之与房东主人的沟通了解后，结合现今及历史，想象并重组，以再现老洋房的原型及建筑功能。为了更好地忠实海派文化，在每幅老洋房的绘图中，我艺术地再现、补充了当年的人文景观，根据环境需要，舍去一些不必要的物体，增加了人物、动物、植物及车辆，营造了画面独特的怀旧氛围，旨在让观者能联想到每一幢花园洋房背后的故事。

13 个月来，工作之余我不停作画，同时，我也在享受、品尝着深入探知老洋房的乐趣。这里大部分图稿是以自然主义的手法逼真再现对象的，而有一小部分图稿是我怀着童话般的梦想，比较主观地以平面化的手法表现对象的。无论哪种艺术手段，我所做的图稿都是在阐述上海滩老洋房的前世今生；以艺术的形式再现和回望历史，感受老上海的百年巨变及神秘的气息。

在这本画册里，读者可以看到史量才曾住过的中西混合式花园住宅、沙逊建造居住的英国式乡村别墅、孙科及其家人一起生活过的西班牙式花园住宅、海上闻人杜月笙的花园公馆，还有藏于虹桥路外墙斑驳的老别墅、铜仁路闻名遐迩的“绿房子”、巨鹿路爬满枝蔓的“爱神花园”。画册里还会遇见毕生繁华、旧上海交际女王唐瑛住宅，上海滩奇女子、锦江红颜董竹君住宅，老上海的骄傲、“国母”宋庆龄故居，上海文化名人、海派作家巴金故居，以及佘山天主教堂、沐恩堂等等。这些从 20 世纪初就矗立在上海的建筑，就像张爱玲小说一样，塑造了“中西合璧”的上海风格。

上海老洋房是历史文化的遗产。原有的花园洋房，有的因是名人故居而修缮开放，如今大部分或恢复为高档住宅，或改建成酒店式公寓。在已修旧如旧的大公馆里，冷不防你会撞见作曲家何训田坐在窗边的椅子上喝着咖啡；在梧桐掩映、叶落缤纷的酒庄草坪上，法国归来的画家方世聪、沪上画家卢象太、作家沈善增、美籍华人画师徐文华等正与一些小众画家进行着沙龙酒会。上海现在的部分区域仍然充满着法式情调的文学艺术气息。如今的老洋房传递的是一种人与人、人与社会的关系，造就着更为优越的人文思想环境和海纳百川的文化品格。它的存在正如同济大学阮仪三教授所言：“建筑是会说话的，它是一种物质形态空间艺术的形象，它本身就富含文化的内涵，并且随着它建筑的形态逐步散发出来，感染后人。”

图已完稿。花园洋房的情调依然留存，但愿我对老上海、老洋房的记忆能为你奉上一份视觉盛宴，共享一次精神之旅途。

胡家康

2014 年 3 月 13 日

Preface

Old buildings commemorate a city in a tangible way, embodying profound history and culture.

Built in the 1920s and 1930s, old exotic and charming Shanghai dwellings in elegant styles and shapes, symbolize the metropolis in unique ways. They are of many classifications: houses of stone, modern high-rise apartment buildings, free-standing ganden villas, row housings,row houses, new-style lane and alley dwellings, modern apartments, homes with verandas, free-standing villas with garths. Shanghai deserves the name "Museum of Historic Buildings".

Born in Shanghai, I spent my childhood in a home near Cité Bourgogne Stone House. After growing up, I studied and worked in old Shanghai because I was especially fond of its grace. Western houses there date back to early in the last century; I as an interested artist, I have dedicated myself to documenting this specialty of Shanghai.

So I have presented the "restored" view as close as possible with the help of sketching, photographing, visual inspection, memory, pictures and written records. No matter rain or scorching heat, I insisted on getting a first-hand view and presenting sketches true to the original buildings. Human landscapes of the times are artistically included, unnecessary are objects left out, and people, animals, plants and vehicles added to enliven the background. By doing so, the sketches help the viewers reminisce about the past and the stories behind every garden house.

I have been painting in my leisure for the past 13 months, enjoying and savoring all the pleasure that exploring these old houses offers me. Most of the sketches herein present an objective natural view, while a few of them are subject views like dreams from my childhood.Whatever the method, I am committed to displaying the past and present of old foreign-style buildings in Shanghai, presenting an artistic restored view of history so as to enable the reader to experience the mystery of the breathtaking changes in the old city over the century.

Now the sketches are finished. The appeals of garden houses presented here are to stay. I hope that my memory of the old city and its Western style houses provide you a visual feast and take you with me on a spiritual journey.

Hu Jiakang
March 13, 2014

安福路、长乐路、多伦路等
Anfu Road, Changle Road, Duolun Road, etc.

吴国桢住宅

（安福路 201 号）

这幢花园洋房建于 1922 年。抗日战争胜利后出任上海市市长的吴国桢曾居住于此。他是周恩来的中学同窗，美国普林斯顿大学哲学博士，曾任蒋介石的私人秘书。

Former Residence of Wu Guozhen

After China won the Anti-Japanese War in 1945, Wu Guozhen served as mayor of Shanghai.
Address: No. 201, Anfu Road.

豫园

（安仁街 132 号）

明代潘允端为奉养其父而于嘉靖三十八年（1559）兴建“豫园”。园中亭、台、楼、阁、假山、池塘一应俱全，充分展现了中国古典园林的建筑风格。

Yu Garden

This garden fully demonstrates the architectural style and design of the Chinese classical gardens.
Address: No.132, Anren Street.

安亭路西班牙式花园住宅

（安亭路 46 号）

该建筑采用了对称式构图，转角塔楼颇具特色，于 1936 年落成。

Spanish Garden House on Anting Road

The building features a corner tower.

Address: No. 46, Anting Road.

玉佛禅寺

（安远路 170 号）

玉佛禅寺创建于清光绪八年（1882），因寺内的两尊缅甸玉佛而得名。该寺几易其址，于 1918 年在今安远路建新寺，10 年后竣工。

Yufo Temple

Yufo Temple was named forits two jade statues of the Buddha.
Address: No.170, Anyuan Road.

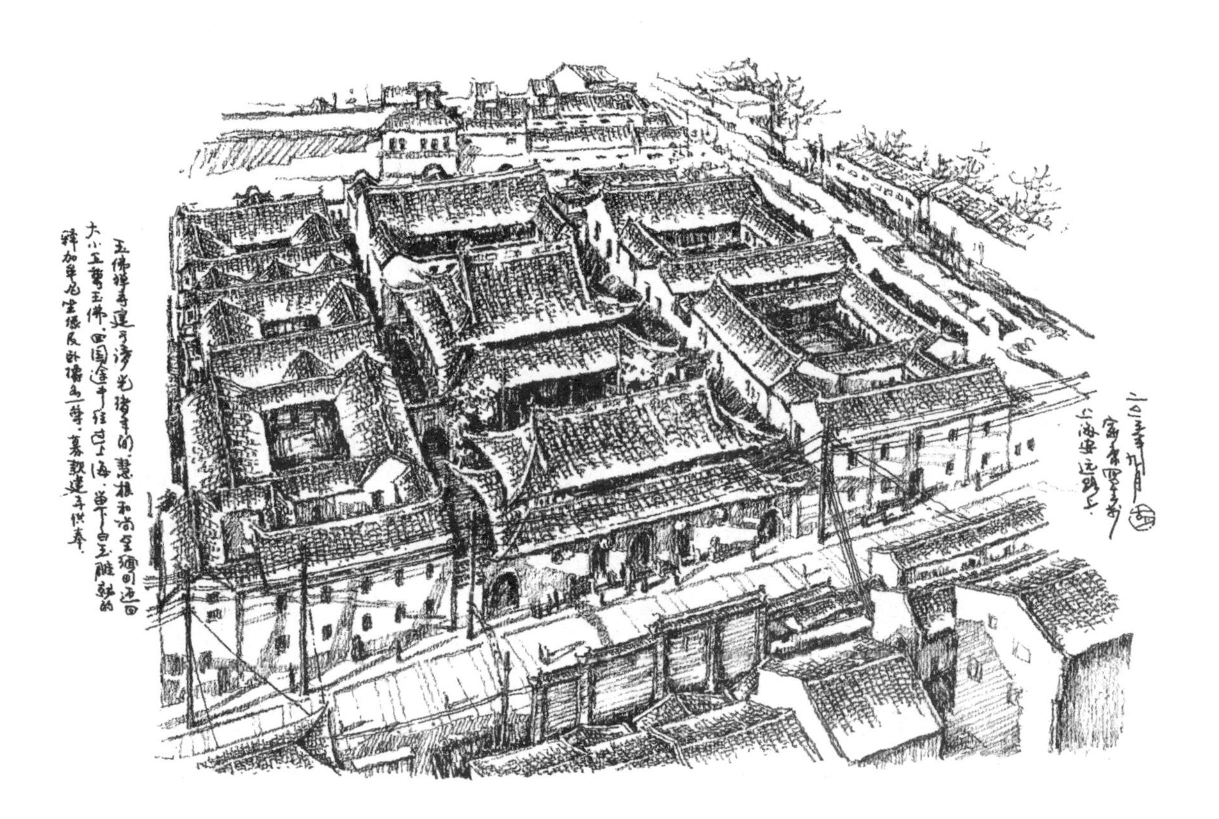

天主教望德堂

（北京西路 1220 弄 2 号）

天主教望德堂竣工于 1932 年，此堂是上海为数不多的由西班牙人自建的建筑，属地道的圣母教堂。

The Augustinian Procuration of the Catholic Church

It is among the few buildings constructed by the Spanish in Shanghai.

Address: No.2, 1220 Lane, West Beijing Road.

爱文公寓

（北京西路 1369 号）

爱文公寓由邬达克设计，1932 年竣工。现今的北京西路在民国时期曾名为爱文义路，该公寓楼因此而得名。

Avenue Apartment

It got its name from Avenue Road.

Address: No.1369, West Beijing Road.

周宗良住宅

（宝庆路 3 号）

有“上海第一私人花园”之称的周宗良住宅，1925 年之前由德国人所建。据说该建筑位居旧上海法租界霞飞路的黄金分割点。

Former Residence of Zhou Zongliang

This is in the “golden section”of Avenue Joffre in the Shanghai French Concession.
Address: No.3, Baoqing Road.

张爱玲居所

（常德路 195 号）

爱林登公寓建于 1936 年。张爱玲曾在这里度过了五年左右的写作时光，完成了《倾城之恋》《沉香屑——第一炉香》《金锁记》《封锁》等名作。

Former Residence of Zhang Ailing

Zhang Ailing, a famous writer, wrote many well known works here over a five-year period.

Address: No.195, Changde Road.

蒲园

（长乐路 570 弄 1—9 号）

蒲园为纪念法国军官蒲石而命名。12 幢西班牙式花园洋房由中国第一代女建筑师张玉泉设计，1942 年竣工。

Garden of Bourgeat

The building was designed by Zhang Yuquan, a female Chinese architect.

Address: No.1-9, 570 Lane, Changle Road.

潘宗周住宅

（长乐路 680 号）

这幢新古典主义花园洋房，建于 20 世纪 20 年代末。该建筑当年的主人潘宗周是旧上海知名商贾，平生喜爱收藏古版书籍，尤爱宋元古版，时为国内屈指可数的“宋版大户”。

Former Residence of Pan Zongzhou

Pan Zongzhou was a well-known collector of ancient books.

Address: No.680, Changle Road.

长乐路古典式花园住宅

（长乐路 800 号）

这幢古典式花园住宅建于 20 世纪 30 年代。建筑底层有敞廊，细卵石墙面，红瓦顶较陡峭，檐下有木支托。

Classic Garden House on Changle Road

It was built in the 1930s.

Address: No.800, Changle Road.

沪西礼拜堂

（长宁路 1465 号）

沪西礼拜堂简称“沪西堂”。这个礼拜堂的前身是由宋庆龄的母亲倪桂珍等人于 1919 年所建立的布道所。

West Shanghai Church

The predecessor of this church was a preaching venue established by Li Guizhen, the mother of Soong Ching-ling, along with others.

Address: No.1465, Changning Road.

摩西会堂

（长阳路 62 号）

摩西会堂于光绪三十三年（1907）始建，1927 年迁至现址。该会堂原仅供犹太人专用，第二次世界大战期间成为犹太难民的宗教活动中心。

Ohel Moishe Synagogue

This was an important venue for Jewish religious activities in Shanghai.

Address: No.62, Changyang Road.

永丰村点式公寓

（重庆南路 177 号、179 弄 1—10 号）

永丰村的多幢公寓竣工于 1919 年。其中南部为点式公寓，北部则大多为毗连式公寓。图中这幢建筑便是诸点式公寓之一。

An apartment in Yongfeng Village

This is one of the apartment in Yongfeng Village.

Address: No.177 and No. 1-10 of 179 Lane, South Chongqing Road.

吕班公寓

（重庆南路 185 号）

这栋公寓建于 1931 年，现名重庆公寓。著名美国女记者艾格尼丝·史沫特莱曾在此从事革命活动。

Lvban Apartment

Agnes Smedley, a famous American reporter, engaged in revolutionary activities here.

Address: No.185, South Chongqing Road.

邹韬奋故居

（重庆南路 205 弄 54 号）

这里原是法租界的上等住宅小区，由法商万国储蓄会投资，90 幢假四层连体建筑共分四行八段，1930 年竣工。近代史上杰出的新闻记者、出版家和政论家邹韬奋租下了 54 号并入住其中。

Former Residence of Zou Taofen

Zou Taofen was a prominent reporter, publisher and politician in modern China. This house is now a memorial hall.

Address: No.54, 205 Lane, South Chongqing Road.

圣伯多禄堂

（重庆南路 270 号）

由法国华侨集资兴建的圣伯多禄堂，始建于 1933 年，次年落成，当时是震旦大学师生专用圣堂。

Saint Peter's Hall

The church served the faculty and students in Aurora University in its early days.

Address: No.270, South Chongqing Road.

大境关帝庙

（大境路 259 号）

大境关帝庙的主殿建在一段明代的老城墙上，非常特别。

Guan Yu Temple on Dajing Road

This unique temple was built on a Ming Dynasty (1368-1644) city wall.

Address: No.259, Dajing Road.

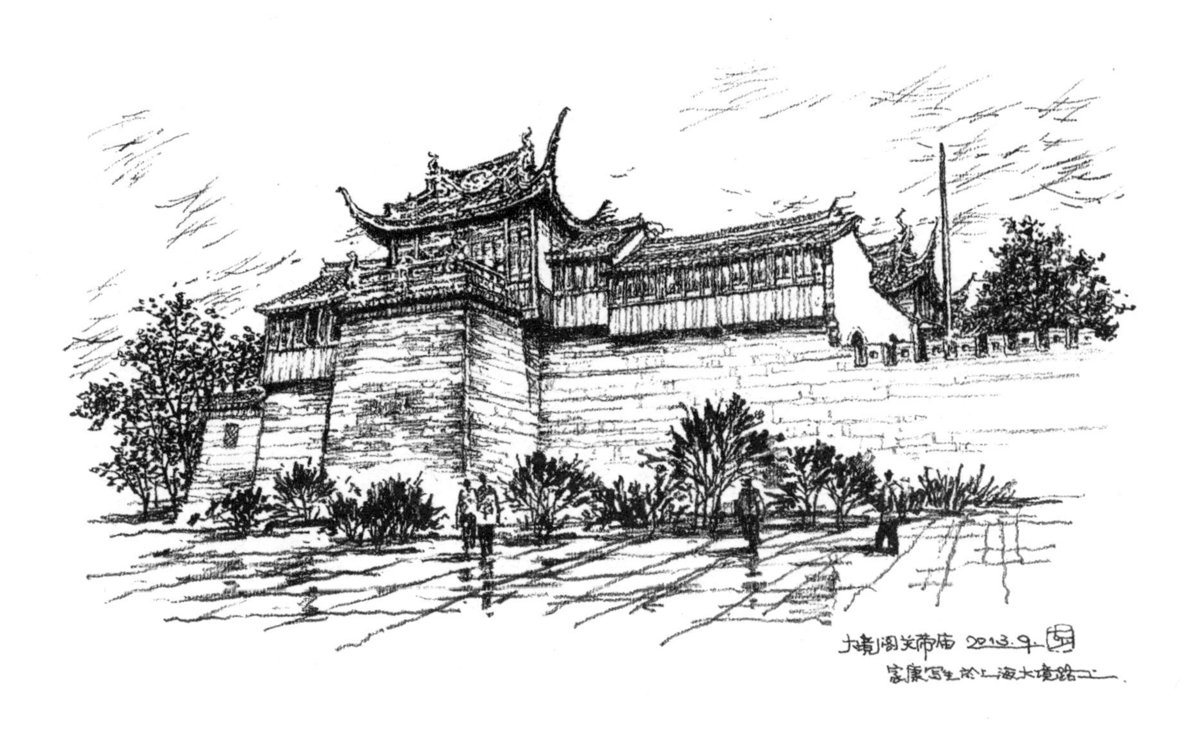

杜公馆

（东湖路 70 号）

据说，金廷荪耗资 30 万美元，于 1934 年建此豪宅送给黑帮大佬杜月笙。然恰逢“8 · 13 事变”爆发，杜氏逃往香港，未曾在此居住。抗战胜利后，杜氏以 60 万美元将此楼卖给了美国新闻处。

Former Residence of Du Yuesheng Mansion

Du Yuesheng was a well-known gangster on the Bund. The building was later sold to the American Press Office.

Address: No.70, Donghu Road.

圣沙忽略堂

（董家渡路 175 号）

原名圣沙勿略天主堂，又称董家渡天主堂，始建于清光绪二十七年（1847）。它是当时远东最大的天主堂，其主保弗朗西斯克·沙勿略是耶稣会派往东亚的第一位传教士。

Saint Xavier Church

Francisco Xavier, patron of this church, was the first Society of Jesus missionaries dispatched to East Asia.

Address: No.175, Dongjiadu Road.

爱庐

（东平路 9 号）

上海音乐学院附中内有数幢法国风格建筑，具体建造时间不详。1927 年蒋宋联姻后，宋子文买下此楼，作为结婚礼物送给了妹妹宋美龄。蒋介石特为之题名“爱庐”。

Ai Lu

Soong Tse-ven bought this house as a wedding gift for Chiang Kai-shek and Soong Mayling.
Address: No.9, Dongping Road.

鸿德堂

（多伦路 59 号）

鸿德堂由美国基督教北长老会与我国信徒共同捐资，于 1928 年建造。这是极少数采用中国古典式建筑风格的教堂。

Fitch Memorial Church

It is one of the few churches of ancient Chinese style.

Address: No. 59, Duolun Road.

李观生住宅

（多伦路 215 号）

这幢西班牙新古典主义风格的建筑原是广东商人李观生的住宅，建于 1924 年。一楼门廊与二楼内阳台采用了古典多立克柱支撑，门前有开阔的庭院。

Former Residence of Li Guansheng

Li Guansheng, a Cantonese merchant, used to live here.

Address: No.215, Duolun Road.

孔祥熙住宅

（多伦路 250 号）

这座阿拉伯风格的建筑是孔祥熙在沪上的三处豪居之一，建造于 1924 年。建筑顶部两座马蹄卷方亭本有圆顶，20 世纪 60 年代被拆除。孔祥熙的政绩有改革中国币制、建设中国的银行体系等。

Former Residence of Kung Hsiang-hsi

Kung Hsiang-hsi contributed to reforming the currency and building a modern banking system in China.

Address: No.250, Duolun Road.

汾阳路、复兴中路、华山路等

Fenyang Road, Middle Fuxing Road, Huashan Road, etc.

木结构独立式花园洋房

（汾阳路9弄3号）

这是上海最老的木结构独立式花园洋房，建于清光绪二十四年（1898），属英国浪漫主义风格。铁皮屋顶，外墙采用横向木板条装饰。

Free-Standing Garden Villa with a Wooden Structure

This is the oldest free-standing wooden garden villa in Shanghai.

Address: No.3, 9 Lane, Fenyang Road.

犹太人俱乐部

（汾阳路 20 号）

1910 年代落成，法国文艺复兴风格。在 20 世纪 30 年代成为了犹太难民的聚集地。

Jewish Club

Many Jewish refugees gathered here in the 1930s.

Address: No.20, Fenyang Road.

丁贵堂住宅

（汾阳路 45 号）

西班牙建筑风格，独立式两层花园小楼，由匈牙利著名建筑师邬达克设计，建于 1932 年。该建筑本系上海江海关为其税务司建造的官邸，前几任主人均属外籍，丁贵堂是入住其中的第一任中国籍长官。

Former Residence of Ding Guitang

It used to be an official mansion of the Inspector General of Jianghai Customs in Shanghai. Several foreigners lived here before Ding Guitang, the first Chinese owner of the house.

Address: No.45, Fenyang Road.

海关副总税务司住宅 <丁贵堂住宅>
2013.7.家农画于上海汾阳路

小白宫

（汾阳路 79 号）

建于清光绪三十一年（1905），因其建筑外形颇似美国白宫，故人称“小白宫”。建国初期，当时的上海市市长陈毅曾居住于此。

The Little White House

It gained this name due to its resemblance of the American White House.

Address: No.79, Fenyang Road.

白公馆

（汾阳路 150 号）

白崇禧一家于 1946 年移居上海，人们把白崇禧、白先勇父子曾经居住过的这座老宅称作“白公馆”。

The Bai Mansion

Bai Chongxi and Bai Xianyong, father and son, lived here.
Address: No.150, Fenyang Road.

麦琪公寓

（复兴西路 24 号）

建于 1937 年，得名于民国时期的麦琪路。该建筑高 40 米，楼前设街心花园，避免了建筑过高所带来的压抑感。

Magy Apartment

It was named after Magy Road during the Republic of China Period (1912-1949).

Address: No.24, West Fuxing Road.

复兴西路英国乡村式住宅

（复兴西路 193 号）

这幢英国乡村式建筑建于 1930 年，有较大的双坡屋顶，红砖砌筑的烟囱，北立面有红色半露木构架，南立面是浅黄色拉毛墙面。

English Country-Style Residence on West Fuxing Road

This red wooden building is unique.

Address: No.193, West Fuxing Road.

复兴西路花园住宅

（复兴西路 193 号）

这栋花园住宅建于 1930 年，属英国建筑风格，位于当年法租界的重要地段。

Garden House on West Fuxing Road

The building was located in an important section of the French Concession in Shanghai. Address: No.193, West Fuxing Road.

诸圣堂

（复兴中路 425 号）

诸圣堂由美国圣公会传教士麦甘霖牧师和中国教师汪孝奎共同筹划，1925 年买地建造。

All Saint's Church

This is a protestant church.

Address: No.425, Middle Fuxing Road.

柳亚子旧居

（复兴中路 517 号）

这幢法国现代式花园住宅建成于 1926 年，是著名诗人、南社创办人、近代民主革命家柳亚子的寓所。柳亚子曾任孙中山临时大总统秘书，新中国成立后曾任全国人大常委会委员。

Former Residence of Liu Yazi

Liu Yazi was a well-known poet, founder of South Society, and a modern China democratic revolutionary.

Address: No.517, Middle Fuxing Road.

克莱门公寓

（复兴中路 1363 弄）

于 1929 年由比利时商人克莱门投资建造，属法国式公寓。当年建筑内每一房间均陈设有法式家具供租赁人使用。

Kremen Apartment

This French-style apartment was funded by Kremen, a Belgian merchant.
Address: 1363 Lane, Middle Fuxing Road.

阿麦伦公寓

（高安路 14 号）

现称高安公寓，建于 1941 年，为现代派风格。

Amyron Apartments

This is a modern architecture.

Address: No.14, Gao'an Road.

荣德生旧居

（高安路 18 弄 20 号）

民国时期著名民族工商业家…… 两根仿意大利文艺复兴时期的多立克廊柱，使建筑品……

Former Residence of Rong ……

Rong Desheng is a well-know…… d (1912-1949).
Address: No.20, 18 Lane, Go……

故园画忆系列
Memory of the Old Home in Sketches

上海老洋房
Old Foreign-style Buildings in Shanghai

學苑出版社
Academy Press

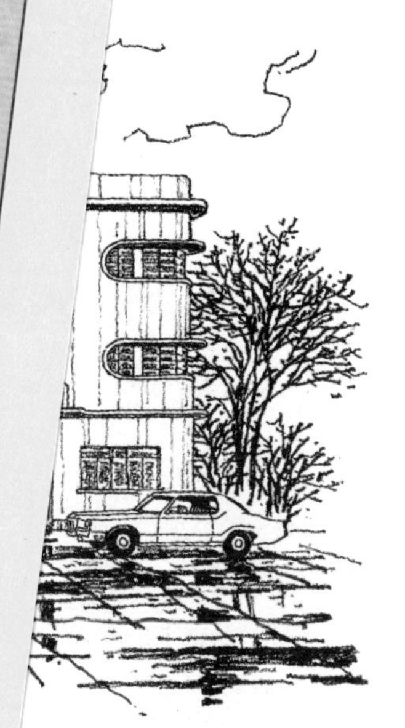

张学良公馆

（皋兰路 1 号）

西班牙式花园洋房，建于 20 世纪 30 年代中期。张学良第三次到上海时居住于此，赵四小姐从北平南来之后亦居于此，因而这里被人们称为“张学良公馆”。

Mansion of Zhang Xueliang

Zhang Xueliang lived here when he came to Shanghai for the third time; Ms. Zhao Si also stayed here after she left Peiping (now Beijing).

Address: No.1, Gaolan Road.

圣尼古拉教堂

（皋兰路 16 号）

1932 年，格列博夫中将在上海俄侨和其他外侨中发起募捐，建造这座东正教大教堂，1934 年落成。教堂建筑属拜占庭式风格。

Saint Nicholas Church

This is a Byzantinesque Eastern Orthodox church.

Address: No.16, Gaolan Road.

沪东礼拜堂

（国和路 350 号）

简称“沪东堂”，前身为“基督复临安息日会中华总会会堂”。随着信徒不断增多，1996 年复建新堂于此，次年落成。

East Shanghai Church

It is the most important church in eastern Shanghai.

Address: No.350, Guohe Road.

国际礼拜堂

（衡山路 53 号）

这是当年上海规模最大的基督教堂，由美侨及其他外国侨民集资建造。其主体为德国仿哥特式教堂建筑，建于 1925 年，初名协和礼拜堂。

Shanghai Community Church

It is the largest Christian church in Shanghai.
Address: No.53, Hengshan Road.

虹桥路英式花园别墅

（虹桥路 2275 号）

这幢英式花园别墅，是虹桥路上建造于 20 世纪 30 至 40 年代系列别墅之一。该建筑的原居民和建造者均已无从查考。

British-Style Garden House on Hongqiao Road

The original builder and resident are unknown.

Address: No. 2275, Hongqiao Road.

沙逊别墅

（虹桥路 2419 号）

沙逊别墅建于 1932 年。原主人维克多·沙逊是 20 世纪初上海著名的房地产大王。

Sassoon Villa

Englishman Victor Sassoon was a well-known real estate giant in Shanghai in the early 20th century.
Address: No.2419, Hongqiao Road.

贺子珍旧居

（湖南路 262 号）

这是一幢建于 1931 年的欧洲独立式花园住宅。原主人是英国锦隆洋行的大股东。汪伪时期该建筑被大汉奸周佛海占据。1954 年，贺子珍从苏联回国后寓居于此，直至 1984 年去世。

Former Residence of He Zizhen

He Zizhen lived here after his return from the Soviet Union until his death.

Address: No.262, Hunan Road.

熊佛西楼

（华山路 630 号）

熊佛西楼名字的由来是为了纪念上海戏剧学院的第一位校长——熊佛西先生。他是现代著名剧作家，中国新兴话剧运动的开拓者之一。

Xiong Foxi Building

This building was dedicated to the first principal of the Shanghai Theater Academy (STA) Mr. Xiong Foxi, a famous playwright and pioneer of newly-developing drama in China.

Address: No.630, Huashan Road.

孙家花园

（华山路 831 号）

这座西班牙风格的三层花园住宅建于 1918 年。据悉，孙家花园原系丁香花园主人李经迈的房产，后来成为“面粉大王”孙多森、孙多鑫兄弟的住宅。

Sun Garden

This was one of the properties of Li Jingmai, owner of the Lilac Garden. It later became the residence of brothers Sun Duosen and Sun Duoxin.

Address: No.831, Huashan Road.

丁香花园

（华山路 849 号）

上海第一座西式花园，始建于清同治元年（1862）。园内三幢别墅均为美国乡村式，是晚清北洋大臣李鸿章为七姨太丁香所建。

The Lilac Garden

Built by Li Hungchang for his seventh wife in the late Qing Dynasty (1644-1911), the Lilac Garden was the first western style garden in Shanghai.

Address: No.849-879, Huashan Road.

郭棣活住宅

（华山路 893 号）

建于 1948 年。为追求新潮的几何造型，平面采用了不对称布局，形体凹凸交错，显示出现代式建筑风格的特征。其主人郭棣活是当年著名的永安集团郭氏家族第二代传人。

Former Residence of Guo Dihuo

This is the mansion of Guo Dihuo, the inheritor of the Guo Family fortune, gained from their profitable business dealings.

Address: No. 893, Huashan Road.

丁香别墅

（华山路 922 号）

这幢法式砖木结构建筑便是著名的丁香别墅。

Lilac Villa

This is a French concrete and wooden structure.

Address: No.922, Huashan Road.

嘉色喇住宅

（华山路 1076 号）

嘉色喇的住宅建造于 1930 年，是一栋典型的德国式花园住宅。

Former Residence of Leopold Cassella

Built in 1930, this house belonged to Cassella, a German merchant.

Address: No.1076, Huashan Road.

华亭路、淮海中路、巨鹿路等

Huating Road, Middle Huaihai Road, Julu Road, etc.

华亭路地中海式花园住宅

（华亭路 71 弄 1 号）

华亭路和延庆路一带的这类建筑多建于 1934 年，由中国建业地产公司建造。这幢建筑属地中海风格。

Mediterranean Garden House on Huating Road

Buildings on Huating and Yan'an Road were built by Central China Real Estate Limited.

Address: No.1, 71 Lane, Huating Road.

华亭路英式花园住宅

（华亭路 71 弄 2—7 号）

华亭路的几幢别墅都建造于 1934 年，除了 1 号楼是地中海风格，其他都是英式风格。

British Garden House on Huating Road

Villas on Huating Road built in 1934 were in British style, except for the first one, in Mediterranean style.
Address: No. 1-7, 71 Lane, Huating Road.

佛兰克林住宅

（淮海西路 338 号）

这幢美国南方庄园式古典花园洋房竣工于 1931 年。此建筑是美籍建筑师哈沙德的代表作品之一。

The Former Franklin Residence

This is a masterpiece of an American architect named Hazzard.

Address: No.338, West Huaihai Road.

苏联驻沪商务代办处

（淮海中路 1110 号）

法国文艺复兴式花园住宅，竣工于 1925 年。20 世纪 50 年代时苏联驻沪商务代办处入驻，今为东湖宾馆 7 号楼。

Former Commercial Agency of the Soviet Union in Shanghai

Formerly housing the Commercial Agency of the Former Soviet Union in Shanghai, it is now the 7th Building, Donghu Hotel.

Address: No.1110, Middle Huaihai Road.

席宅

（淮海中路 1131 号）

仿古典德式建筑，建于 1926 年，其主人席正甫曾是汇丰银行的老板。德式建筑在上海并不多见，因二战结束后一大批德国建筑被拆除。

Former Residence of Xi Zhengfu

A rare German-style building in Shanghai, it belonged to Xi Zhengfu, formerly in charge of the Hong Kong and Shanghai Banking Corporation (HSBC).

Address: No.1131, Middle Huaihai Road.

赵丹故居

（淮海中路 1273 弄）

新康花园建于 1934 年，共有 11 幢二层建筑，每幢内皆分上、下两套住宅。此类格局在近代上海花园住宅中并不多见。著名电影表演艺术家赵丹曾在此居住。

Former Residence of Zhao Dan

Zhao Dan was a famous film actor.

Address: 1273 Lane, Middle Huaihai Road.

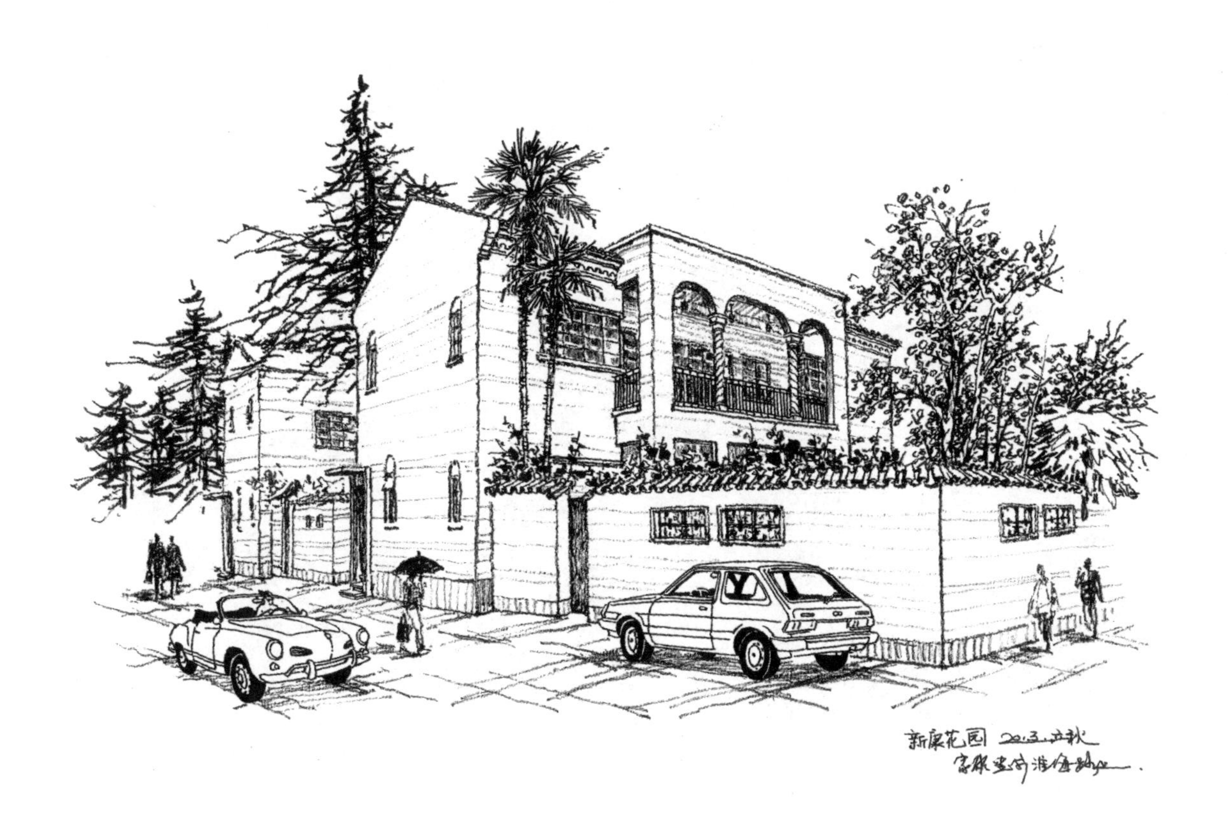

上方花园

（淮海中路 1285 弄）

原名沙发花园。1933 年浙江商业银行向英籍犹太人沙发购园，1938−1941 年分批建造完成。

Shangfang Garden

Originally named Sofa Garden, it was owned by a British Jew.

Address: 1285 Lane, Middle Huaihai Road.

巴塞住宅

（淮海中路 1431 号）

这幢西班牙与意大利风格兼融的建筑造于 1921 年，原为巴塞住宅，后相继被作为西班牙、朝鲜、法国等国的领事馆使用。

The Former Basset Residence

It has in succession been the consulate for Spain, North Korea and France.

Address: No. 1431, Middle Huaihai Road.

盛宣怀住宅

（淮海中路 1517 号）

新古典主义风格的洋房，建于清光绪二十六年（1900），后被清代洋务派主要人物之一的盛宣怀购得，于是被称为盛宣怀住宅。抗战期间，日本人占用了此建筑，并将其花园一半辟建为如今的上海新村。

Former Residence of Sheng Xuanhuai

Sheng Xuanhuai was an important advocate of the westernization movement during the Qing Dynasty (1644-1911). This foreign-style building was seized by the government during the Republic of China (1912-1949), and then occupied by the Japanese forces in the Anti-Japanese War (1937-1945).

Address: No.1517, Middle Huaihai Road.

逸邨

（淮海中路 1610 弄 2 号）

这些西班牙花园住宅建于 1942 年。蒋经国一家曾居住于此。蒋氏在逸邨居住的最后三个月里写下了《沪滨日记》。

Yi Cun

The family of Chiang Ching-kuo, a son of Chiang Kai-shek used to live here.

Address: No.2,1610 Lane, Huaihai Middle Road.

何应钦住宅

（淮海中路 1634 号）

何应钦的这栋住宅建于 1930 年。他于清宣统元年（1909）在日本加入中国同盟会，1911 年参加辛亥革命，后担任国民党行政院国防部长、行政院院长等要职，1945 年代表中国政府接受日本投降。

Former Residence of He Yingqin

A significant politician in modern China, He Yingqin accepted Japan's surrender on behalf of the Chinese government in 1945.

Address: No. 1634, Middle Huaihai Road.

淮海中路花园里弄住宅

（淮海中路 1754 弄）

这是建于 20 世纪 30 年代的西班牙风格建筑群。

Lane-and-Alley Garden Houses on Middle Huaihai Road

This Spanish style complex was built in the 1930s.

Address: 1754 Lane, Middle Huaihai Road.

诺曼底公寓

（淮海中路 1842—1858 号）

建于 1924 年，由法商万国储蓄会投资，邬达克设计，属法国文艺复兴建筑风格。它是沪上最早的外廊式高层公寓。

Normandy Apartment

Normandy Apartment is the first balcony-style highrise in Shanghai.

Address: No. 1842-1858, Middle Huaihai Road.

宋庆龄故居

（淮海中路 1843 号）

这栋住宅建于 20 世纪 20 年代初。这是宋庆龄一生中居住时间最长的地方，她曾在这里会见过众多国家政要。

Former Residence of Soong Ching-ling

This is where Soong Ching-ling lived for the longest time and met with state leaders.

Address: No.1843, Middle Huaihai Road.

贝家老宅

（黄陂南路 25 号）

这幢欧洲古典风格的建筑建于清宣统二年（1910），是清末贝润生的旧宅，在 2001 年开辟延安中路绿地工程时被拆除。

The Old Bei Family House

The former residence of Bei Runsheng in the late Qing Dynasty (1644-1911), it was torn down due to Yan'an Middle Road Green Project in 2001.

Address: No.25, South Huangpi Road.

甘村新式里弄住宅

（嘉善路 131—143 弄、169 弄）

这类住宅曾是上海中产阶级的主要聚集区。

Newly-Developing Lanes and Alleys, Gancun Village

The Bourgeoisie in Shanghai used to live in this type residence.

Address: 131-143 Lanes and 169 Lane, Jiashan Road.

法国太子公寓

（建国西路 394 号）

这幢公寓又名道雯南公寓，约竣工于 1935 年，现代派建筑风格。设计者为法籍建筑师事务所赉安洋行。

Former Residence of a French Prince

This residence was built by French architectural firm.

Address: No.394, West Jianguo Road.

懿园

（建国西路 506 弄）

建于 1941 年的这组新式里弄住宅，具西班牙式和英国乡村式两种风格。这里曾是民国政府财政部官员邹琳的旧居。

Yi Garden

This is the former residence of Zou Lin, an official serving in the Ministry of Finance during the Republic of China Period (1912-1949).

Address: 506 Lane, West Jianguo Road.

王时新住宅

（建国西路 628 号）

这幢现代花园住宅落成于 1948 年，原为香港富豪王时新的故居。

Former Residence of Wang Shixin

This garden house formerly belonged to Wang Shixin, a Hong Kong billionaire.

Address: No. 628, West Jianguo Road.

圣三一基督教堂

（九江路 219 号）

这是上海现存最早的基督教新教英国圣公会主教座堂，俗称“红礼拜堂”。

Holy Trinity Cathedral

This is the oldest Protestant Missionary Society church.

Address: No.219, Jiujiang Road.

刘吉生故居

（巨鹿路 675−681 号）

这栋住宅原名“爱神花园”，是近代著名实业家刘吉生的故居，由匈牙利建筑师邬达克设计，建于 1931 年，属仿古典建筑风格。

Former Residence of Liu Jisheng

Liu Jisheng was an eminent modern industrialist, his former house now the headquarters of the Shanghai Writers' Association.

Address: No.675-681, Julu Road.

巨鹿路英国乡村式花园住宅

（巨鹿路 852 弄 1—8 号、10 号）

这座英国乡村式花园住宅建于 1930 年。其当初的主人已无从查考。建筑为砖木结构，分列弄堂两侧。黄色水泥砂浆的粗糙墙面和那些高耸的烟囱，常常令观者过目难忘。

Garden House on Julu Road

This English-style building's yellow walls and tall chimneys make a deep impression on visitors.
Address: No.1-8 and No.10, 852 Lane, Julu Road.

巨鹿路英式双毗连花园住宅

（巨鹿路 889 号）

这幢英式双毗连花园住宅由当年的亚西亚火油公司建造，1929 年竣工，位于巨鹿路 889 号。与之同时建造的共有 9 幢。

British-Style Double Garden House on Julu Road

This house formerly accommodated senior staff of the Asiatic Petroleum Company. There are 9 such houses here.

Address: No. 889, Julu Road.

康平路花园住宅

（康平路 1 号）

这幢花园住宅建于 1930 年。哲学家、教育家周抗和实业家、大收藏家李祖夔都曾居住于此。

Garden House on Kangping Road

Zhou Kang, a great philosopher and educator, and Li Zukui, an eminent industrialist and collector, once lived here.

Address: No.1, Kangping Road.

瑞金二路、陕西北路、铜仁路等

2nd Ruijin Road, North Shaanxi Road, Tongren Road, etc.

焉息堂

（可乐路 1 号）

焉息堂，亦即今西郊天主教堂，是上海罕见的座拜占庭风格的教堂，始建于 1925 年。

Catholic Country Church

This is a rare Byzantinesque cathedral in Shanghai.

Address: No.1, Kele Road.

景灵堂

（昆山路 135 号）

景灵堂原名“景林堂”，由美南监理会传教士、中西书院创始人林乐知创建。宋耀如全家均为其信徒，并在此堂为蒋介石施行洗礼。

Jingling Church

Jingling Church was the second church American Southern Methodist Episcopal Mission built in Shanghai.

Address: No. 135, Kunshan Road.

真如寺

（兰溪路 399 号）

真如寺位于真如镇北首，原名“万寿寺”，俗称“大庙”。该寺创建于南宋嘉定年间。其大殿系全国佛教寺院中为数不多、保存较好的元代建筑，殊为珍贵。

Zhenru Temple

The main hall of Zhenru Temple, a precious relic, is one of few intact Yuan Dynasty (1206-1368) buildings in china.

Address: No.399, Lanxi Road, Zhenru Town.

新天地石库门

（马当路、兴业路）

“新天地”所辖马当路、兴业路一带近代的石库门建筑，是这座城市与众不同的标识。这里的石库门建筑多落成于20世纪20年代至30年代。

Stone Houses near Shanghai New World

Stone houses in the neighborhood were mostly completed in the 1920s and 1930s.

Address: Madang Road and Xingye Road.

龙华古寺

（龙华路 2853 号）

龙华古寺相传始建于三国吴赤乌五年（242），是江南地区最古老的寺庙之一。

Longhua Ancient Temple

This is one of the oldest temples south of the Yangtze River.

Address: No.2853, Longhua Road.

清心堂

（陆家浜路 675 号）

清心堂原是美国人创建的上海长老会第一会堂，信徒多为社会底层的劳动人民。

Pure Heart Church

Built by the Americans, it is the First Hall of Presbyterian Church in Shanghai.
Address: No.675, Lujiabang Road.

新天安堂

（南苏州路 107 号）

新天安堂亦名联合教堂，建于清光绪十二年（1886）。这是一座废弃已久的老教堂，其所处的位置正是上海近代城市发展的起点。

The Union Church

Having been long abandoned, it stands where modern Shanghai began its development.
Address: No.107, South Suzhou Road.

徐家汇天主堂

（蒲西路 158 号）

徐家汇天主堂，原名为圣依纳爵主教座堂，清宣统二年（1910）竣工。该教堂系典型的法国哥特式双尖顶砖石结构建筑，是上海最大的天主教堂。

Xujiahui Catholic Church

It is the biggest catholic church in Shanghai, showing a typical French catholic style of masonry structure with double pinnacles.

Address: No.158, Puxi Road.

周湘云住宅

（青海路 44 号）

周湘云住宅竣工于 1936 年，新瑞和洋行设计，现代式风格。此宅布局严谨，建筑平面设计工整，造型和谐。屋主周湘云为当年上海的房地产巨商。

Former Residence of Zhou Xiangyun

Zhou Xiangyun was a real estate developer in modern Shanghai.

Address: No.44, Qinghai Road.

吴妙生住宅

（钱仓路 316 号）

吴妙生的这座江南四合院式民居建于 1924 年。砖木结构，正面朝南，二进一院，四周外墙均砌清水砖嵌红砖带饰，立面采用西方古典装饰。此宅于 2009 年为迎接上海世博会整修浦东大道时被拆除。

Former Residence of Wu Miaosheng

This residence was torn down when Pudong Road was renovated for the Shanghai Expo in 2009.
Address: No.316, Qiancang Road, Pudong New Area.

四明公所

（人民路 852 号）

“四明公所”建于清嘉庆二年（1797），是旅沪宁波商人和手工业者的行会组织。

Siming Association

The Siming Association used to be a guild of Ningbo businessmen and handicraftsmen in Shanghai.

Address: No. 852, Renmin Road.

英商马立斯住宅

（瑞金二路 118 号）

瑞金宾馆 1 号楼是英国商人马立斯的私人花园别墅之一，建于 1917 年。

Former Residence of Maris

It used to be one of the private garden villas of Maris, an English businessman.
Address: No.118, 2nd Ruijin Road.

三井洋行大班住宅

（瑞金二路 118 号）

瑞金宾馆 4 号楼原为三井洋行大班住宅，1924 年竣工。建筑为法国古典式花园住宅。

Residence for the Senior Group of the Mitsui Bussan Kaisha Co., Ltd

This was the residence for the senior group of the Mitsui Bussan Kaisha Co., Ltd.
Address: No.118, 2nd Ruijin Road.

托益住宅

（陕西北路 80 号）

这幢独立式花园洋房是英国商人托益于清光绪二十六年为自己居住而投资建造的，解放后曾为上海第二工业大学使用。该建筑于 20 世纪 90 年代被部分拆除。

Former Residence of Toy

This was built by Toy, a British merchant, and later used by the Shanghai Second Polytechnic University after the 1949 founding of the People's Republic of China.

Address: No.80, North Shaanxi Road.

怀恩堂

（陕西北路 375 号）

怀恩堂建成于 1942 年，是上海各基督教会可容纳人数最多的教堂之一。

Grace Baptist Church

It is one of the Christian churches with the largest capacity in Shanghai.

Address: No.375, North Shaanxi Road.

西摩会堂

（陕西北路 500 号）

西摩会堂，又名拉希尔会堂，1920 年落成，属希腊神殿式建筑。这是远东地区现存最大的犹太教会堂。

Ohel Rachel Synagogue

It is the largest existing synagogue in the Far East.

Address: No.500, North Shaanxi Road.

步高里

（陕西南路 287 号）

从这个弄堂口进去，是一片法式联排式的石库门建筑，建造于 1930 年。

Cité Bourgogne

Along the lane are adjoining Stone Houses in the French style.

Address: No. 287, South Shaanxi Road.

若瑟堂

（四川南路 36 号）

若瑟堂又名圣约瑟教堂，是上海较早的天主教堂。清咸丰十年（1860）动工，次年竣工，因位于当年的洋泾浜（今延安东路）河道南岸，而又被称为洋泾浜天主堂。

St. Josephis Church

This is one of the early catholic churches in Shanghai.

Address: No. 36, South Sichuan Road.

袁佐良寓所

（思南路 41 号）

这栋西班牙风格的独立式花园洋房，建于 20 世纪 20 年代。据传说，这栋洋房的主人是袁佐良，一位 20 世纪 20 至 40 年代的金融界名人。

Property of Legendary Yuan Zuoliang

It is said that this garden house belonged to Yuan Zuoliang, a celebrity in the financial community in the 1920s to 1940s.

Address: No.41, Sinan Road.

思南公馆别墅群

（思南路 51—95 号）

由 22 幢别墅构成的思南公馆，原名“义品村”，于 1932 年至 1936 年由比利时义品洋行投资建造。周恩来、梅兰芳、罗隆基等许多知名人士曾在这一带居住过。

Sinan Mansion

Many celebrities, including Zhou Enlai, Mei Lanfang, and Luo Longji, once lived in this neighborhood.

Address: No.51-95, Sinan Road.

贺绿汀旧居

（泰安路 76 弄 4 号）

这里共有 14 幢独立的西式住宅，建造于 1934 年。著名音乐家贺绿汀的大半生时间均在此居住，创作了许多广为人知的作品。

Former Residence of He Lvting

He Lvting was a famous musician; his compositions are popular among the public.

Address: No.4, 76 Lane, Tai'an Road.

陈家巷乡村

（泰安路 115 弄）

这些英国古典式和西班牙式假三层洋房，由黄迈士设计，德士古洋行建造，于 1948 年竣工，作为德士古洋行高级职员住宅。1949 年解放前夕，外侨纷纷撤离回国。

Chenjiaxiang Village

These buildings were constructed by Texaco Bank to accommodate senior staff.

Address: 115 Lane, Tai'an Road.

卫乐园

（泰安路 120 弄）

卫乐园始建于 1924 年，当年多为金融界上层人士租用。每幢建筑和格局皆有不同，多属西班牙式和英国乡村式。

Willow Garden

These apartments were mainly for senior financiers.
Address: 120 Lane, Tai'an Road.

马歇尔公寓

（太原路 160 号）

又名太原别墅，建于 1928 年，属法国晚期文艺复兴风格的花园住宅。这里曾接待过诸多国家领导人和外国元首。

Marshall Mansion

Many Chinese top leaders and heads of state from abroad visited here.

Address: No.160, Taiyuan Road.

露德圣母堂

（唐镇街 40 号）

这座教堂仿法国露德圣母大殿式样而造，是 19 世纪中叶浦东地区的传教中心。

Our Lady of Lourdes Church

This was a missionary center in the Pudong area in the middle of the 19th century.

Address: No.40, Tangzhen Street, Pudong New Area.

张叔驯住宅

（天平路 40 号）

这幢西班牙风格与日本庭园风格相结合的建筑建于 1943 年。其本为古钱币收藏家张叔驯的住宅，后为民国上海市市长机要秘书之私宅。

Former Residence of Zhang Shuxun

Zhang Shuxun was a famous collector of ancient coins in modern China. This house later became the home of the confidential secretary of the mayor of Shanghai during the Republic of China Period (1912-1949).

Address: No.40, Tianping Road.

史量才旧居

（铜仁路 257 号）

这是近代杰出新闻事业家、著名爱国民主人士史量才的旧居，建于 1922 年。史量才着眼社会事业，以兴国为己任，将新闻作为文化进步之先锋，使《申报》发扬光大。

Former Residence of Shi Liangcai

Shi Liangcai was a patriot democrat and director of Shun Pao(an influential newspaper started in 1872).

Address: No.257, Tongren Road.

铜仁路毗连式公寓

（铜仁路 280 号）

此公寓建于 20 世纪 30 年代，属砖混结构，前有公共庭院，北立面的四个三角形山墙呈对称格局，南立面的大阳台间隔地凸出。

Adjoining Apartments on Tongren Road

These apartments share a courtyard.
Address: No.280, Tongren Road.

吴同文花园住宅

（铜仁路 333 号）

煤油商吴同文的花园住宅始建于 1936 年，1938 年竣工。此宅由邬达克设计，主立面颇似豪华邮轮，属现代派花园风格。又因建筑外表贴有绿色釉面砖而被称为“绿房子”。

Garden House of Wu Tongwen

This Garden House is decorated with green glaze tile, hence the name “Green House”.

Address: No.333, Tongren Road.

邱氏住宅

（威海路 412 号）

这幢欧洲城堡式花园住宅建于 20 世纪 20 年代，原宅主人是邱倍山、邱渭卿兄弟，上海滩四大染料商之一。两幢相似的豪宅，称为东楼和西楼。图中是东楼，1940 年被民立中学买下，使用至今。

Qiu Mansion

This is the east wing of the house formerly owned by the Qiu brothers. It was purchased by the Shanghai Minli Middle School and has been in service ever since.

Address: No.412, Weihai Road.

武康路、新华路、兴国路等

Wukang Road, Xinhua Road, Xingguo Road, etc.

正广和洋行大班住宅

（武康路 99 号）

这幢建筑建于 1928 年，原是英国正广和洋行的产业。该洋行的汽水曾是上海家喻户晓的饮料。

Former Residence for the Senior Group of Aquarium Bank

This building formerly belonged to the English Aquarius Bank, a company that mainly sold a soda known to every household in Shanghai.

Address: No.99, Wukang Road.

巴金故居

（武康路 113 号）

这栋老建筑是著名作家巴金在上海居住最久的地方，建于 1932 年。

Residence of Ba Jin

Ba Jin, a famous writer, lived here for the longest time in Shanghai.
Address: No.113, Wukang Road.

原意大利总领事官邸

（武康路 390 号）

这栋地中海式花园建筑是原意大利总领事的官邸。

Former Italian Consulate General

This Mediterranean garden house was the office building of the former Italian Consulate General.

Address: No.390, Wukang Road.

黄兴故居

（武康路 393 号）

这幢古典主义装饰派的建筑竣工于 1915 年。这曾是中国近代著名民主革命家黄兴的故居，后又被国民党四大元老之一的李石作为世界社社址。

Former Residence of Huang Xing

Huang Xing was a famous revolutionary in modern China. Later Li Shi based the office of the World Society Magazine here.

Address: No.393, Wukang Road.

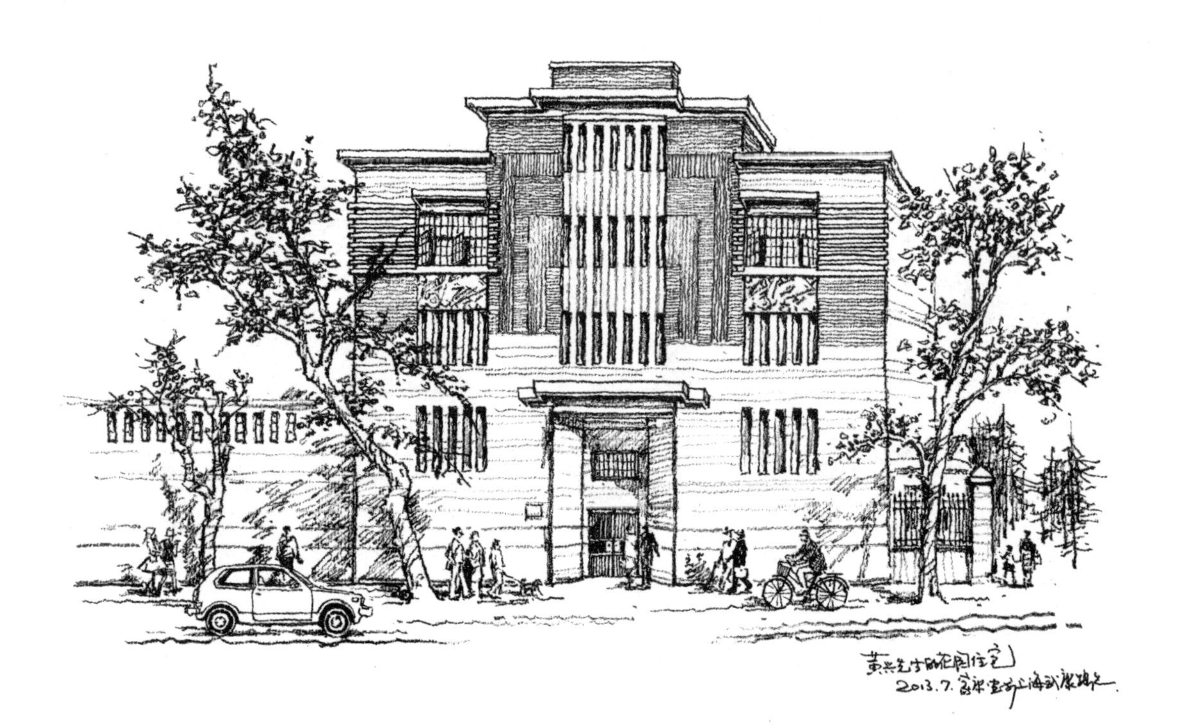

朱敏堂住宅

（乌鲁木齐南路 151 号）

这栋美国复兴式花园建筑当年是大统织染厂经理朱敏堂的住宅，1948 年竣工。此宅于 20 世纪 90 年代中期被拆除，后依原样重建，其中门廊等构件为原物。

Former Residence of Zhu Mintang

Torn down in the mid-1990s, the villa was later rebuilt to its original appearance.

Address: No.151, South Urumchi Road.

东方汇理银行大班故居

（吴兴路 87 号）

这幢建于 1928 年的现代式风格独立花园住宅，是著名的丽波花园的一部分，曾为法国东方汇理银行大班的寓所。

Former Residence of the Senior Group of the Banque de L'indochine

This morden free-standing villa is a part of the famous Lipo garden.

Address: No.87, Wuxing Road.

武夷路花园别墅

（武夷路 127 号）

这幢假三层法国式花园别墅建于 1932 年。该建筑体量较大，其原主人 1949 年离开上海后便被收归国有，先后有上海日用化工疗养院、意大利领事馆和比利时领事馆入驻。

Garden Villa on Wuyi Road

This villa served successively as a sanatorium, the Italian Consulate and the Belgian Consulate after the People's Republic of China was founded in 1949.

Address: No.127, Wuyi Road.

沐恩堂

（西藏中路 316 号）

沐恩堂原名慕尔堂，又名“慕乐堂”，属基督教美国卫理斯教派，是当时上海乃至远东地区的著名教堂。

Moore Memorial Church

It was a famous church in Shanghai and the Far East back then.
Address: No.316, Middle Tibet Road.

龚品梅故居

（香山路 6 号）

这幢法国文艺复兴式花园住宅，建造于 20 世纪 20 年代。该建筑曾是天主教原中国上海教区主教龚品梅的住所。

Former Residence of Gong Pinmei

Gong Pinmei was the Bishop of the former Roman Catholic Diocese of Shanghai, China.
Address: No.6, Xiangshan Road.

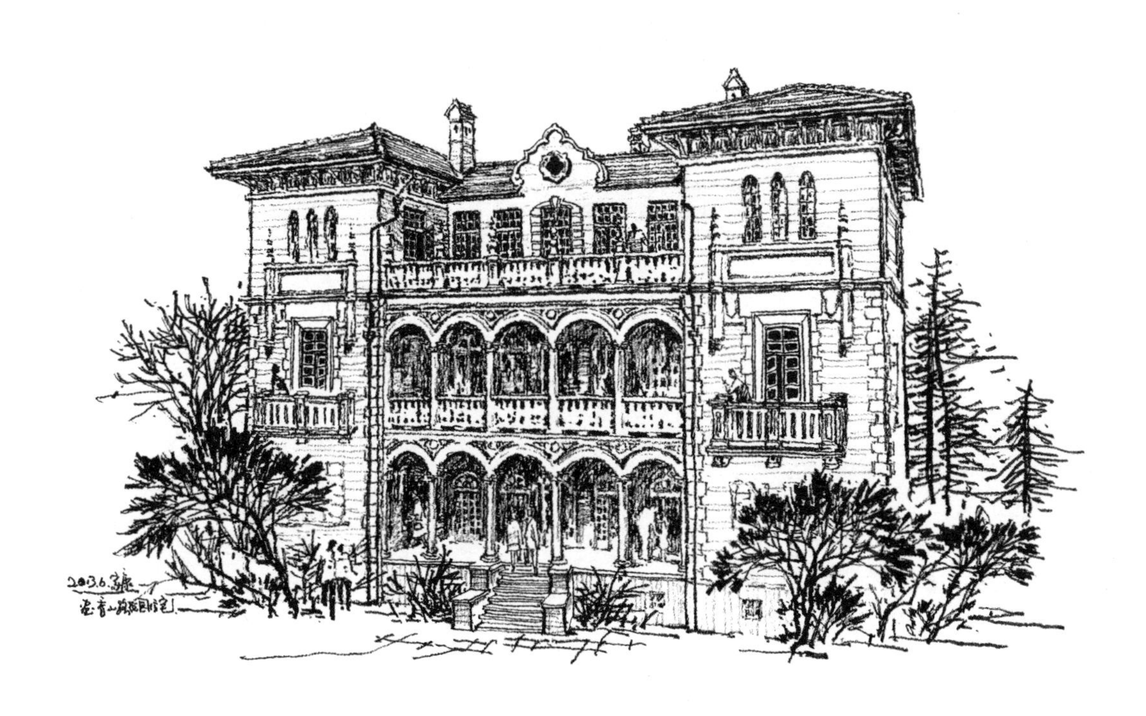

孙中山故居

（香山路7号）

1918年，四名加拿大华侨将这栋欧洲乡村式小洋房赠与孙中山先生。孙中山与宋庆龄曾在此共同居住。1925年孙中山先生逝世后，宋庆龄继续在此居住至1937年。抗日战争爆发后，宋庆龄先后移居香港、重庆，1945年底回至上海，将此建筑移赠国民政府，作为孙中山的永久纪念地。

Former Residence of Sun Yatsen

This is where Sun Yatsen lived for the longest time. The building is now the Dr. Sun Yatsen Memorial Hall of Shanghai.

Address: No.7, Xiangshan Road.

西城回教堂

（小桃园街 52 号）

西城回教堂，亦称清真西寺，是上海穆斯林宗教、文化活动中心。该教堂初建于 1917 年，1925 年在现址重建，是一座有四座圆顶，具有西亚伊斯兰建筑风格的清真寺。

West Shanghai Mosque

This is a center for Muslim religious and cultural activities in Shanghai.

Address: No. 52, Xiaotaoyuan Street.

新华路德式居民别墅

（新华路 179 号）

这栋德国式居民别墅建于 1925 年，现为新华路警署使用。原底层南向有宽敞门廊，今已拆除。建筑上端的咖啡色木梁、木撑不仅有助于结构的柔韧性，还具有十分独特的装饰效果。

German-Style Villa on Xinhua Road

Formerly serving oversea settlers, it is now a police station.

Address: No.179, Xinhua Road.

June 14th, 2013 Jack

李佳白住宅

（新华路 211 弄 2 号）

美国传教士李佳白自清光绪八年（1882）起在华传教长达 40 余年。该建筑于 1925 年竣工使用。建筑的草坪一侧有网球场、小型游泳池等设施，环境极其幽静。

Former Residence Of Li Jiabai

Li Jiabai (AKA Gilbert Reid) was an American missionary, preaching in china for more than 4 decades. This building has a tennis court and small swimming pool.

Address: No.2, 211 Lane, Xinhua Road.

新华路英式乡村别墅

（新华路 236 号）

与此建筑临近的诸别墅均建于 20 世纪 30 年代，其风格除英式乡村别墅外，还有现代式花园别墅。

British Country Style Villa on Xinhua Road

A typical British country style Villa.

Address: No.236, Xinhua Road.

新华路花园住宅

（新华路 315 号）

这幢建于 1930 年的花园住宅系英国乡村风格。据传，该建筑原主人是当年某外籍石油巨商之子。

Garden House on Xinhua Road

It is said the owner was the son of a foreign oil magnate.

Address: No.315, Xinhua Road.

新华路西班牙式花园住宅

（新华路 329 弄 17 号）

这一带有数十幢风格各异的花园别墅，曾经的住户大多数是外国侨民。这幢西班牙式花园住宅竣工于1936 年。

Spanish Garden House on Xinhua Road

This neighborhood has dozens of garden villas of various styles, mainly accommodating overseas Chinese in the past.

Address: No.17, 329 Lane, Xinhua Road.

周均时住宅

（新华路 329 弄 36 号）

这是一幢极罕见的双层圆形花园别墅，被称为“蛋糕房”。该建筑落成于 20 世纪 20 年代末，本属西班牙公使官邸，后为民主人士、原同济大学校长周均时购得。

Former Residence of Zhou Junshi

This is a rare two-story garden villa, thus also called the “Cake House”.
Address: No.36, 329 Lane, Xinhua Road.

梅泉别墅

（新华路 503 弄 1—20 号）

此别墅群由著名建筑师奚福泉设计，建于 1933 年。16、20 号等庭院内设有清澈的水井。小区入口右侧曾有一花园，其中置有两公尺见方的喷泉水池一座，周边梅树错落，以喻“梅泉”。

Meiquan Villas

These villas were designed by Xi Fuquan, a well-known architect.

Address: No.1-20, 503 Lane, Xinhua Road.

东正教圣母大堂

（新乐路 55 号）

此堂建于 1932 年，由东正教上海教区主教维克托尔（白俄）向教徒和白俄侨民集资建造。这是上海现存最完好的东正教教堂。

Russian Orthodox Church

This is the best-preserved Eastern Orthodox church in Shanghai.

Address: No. 55, Xinle Road.

英商太古洋行大班住宅

（兴国路 72 号）

建于 1934 年的这幢太古洋行大班住宅，位于兴国宾馆 1 号楼，属英式皇家古典建筑风格。

Residence for the Senior Group of the Butterfield & Swire Company

This was the residence for the senior group of the Butterfield & Swire Company.
Address: No.72, Xingguo Road.

兴国路英国维多利亚滨海建筑

（兴国路 72 号）

建于 20 世纪 20 年代至 30 年代，位于兴国宾馆 2 号楼，假三层砖木结构，山墙及歇山部分用垂直黑色细木条装饰，非常古朴独特。

British Victorian Buildings in the Coastal Area on Xingguo Road

The unique building has an ancient charm.

Address: No.72, Xingguo Road.

兴国宾馆6号楼

（兴国路 72 号）

这幢法国风格的建筑建于 20 世纪 20 至 30 年代，原是一位美国商人的住宅，后来被商务印书馆的经理和大股东李拔可买下。

Building No.6, Radisson Plaza Xing Guo Hotel Shanghai

This French-style building was the home of an American merchant;it was later purchased by Li Bake, a wealthy Fujian businessman.

Address: No.72, Xingguo Road.

延安西路、永嘉路、愚园路等

West Yan'an Road, Yongjia Road, Yuyuan Road, etc.

孙科住宅

（延安西路 1262 号）

这是孙中山先生之子孙科的住宅，系西班牙式建筑，但兼有巴洛克风格，壁炉顶端的烟囱又似是意大利文艺复兴时期的表现手法。该建筑由邬达克设计，约建于 20 世纪 30 年代。

Former Residence of Sun Ko

Sun Ko was the son of Sun Yatsen. The house was designed by Laszlo Hudec.

Address: No.1262, West Yan'an Road.

延安西路西班牙式花园住宅

（延安西路 2658 号）

该建筑属西班牙花园住宅风格，建于 1934 年。

Spanish Garden House on West Yan'an Road

This elegant garden house is of a Spanish style.
Address: No. 2658, West Yan'an Road.

延安中路英侨住宅

（延安中路810号）

这栋英式建筑建于1920年代，原是英国侨民亚伯拉罕的住宅。20世纪50年代初，他离开上海回英国。

Former Residence of a British-Born Chinese gentleman on Middle Yan'an Road

It used to be the house of Abraham, a British born Chinese who returned to Britain in the 1950s.

Address: No.810, Middle Yan'an Road.

马勒别墅

（延安中路、陕西南路口）

这座挪威式私人城堡的设计源自马勒小女儿的一个梦。历时九年，于 1936 年竣工。1941 年马勒为躲避战乱离开中国，这座“梦幻城堡”却永远留在了上海。

Moller Villa

The design of this castle villa was a dream of Moller's little daughter.
Address: Intersection of Middle Yan'an Road and South Shaanxi Road.

永业大楼

（雁荡路 6 号）

永业大楼原名杨氏公寓，建于 1932 年。当年入住者多为中外高级职员。各单元楼梯通道的顶层为电梯机房和蓄水箱，大楼转弯处顶部设有圆形日光浴室。

Yongye Building

This building was originally named Young Apartments.

Address: No.6, Yandang Road.

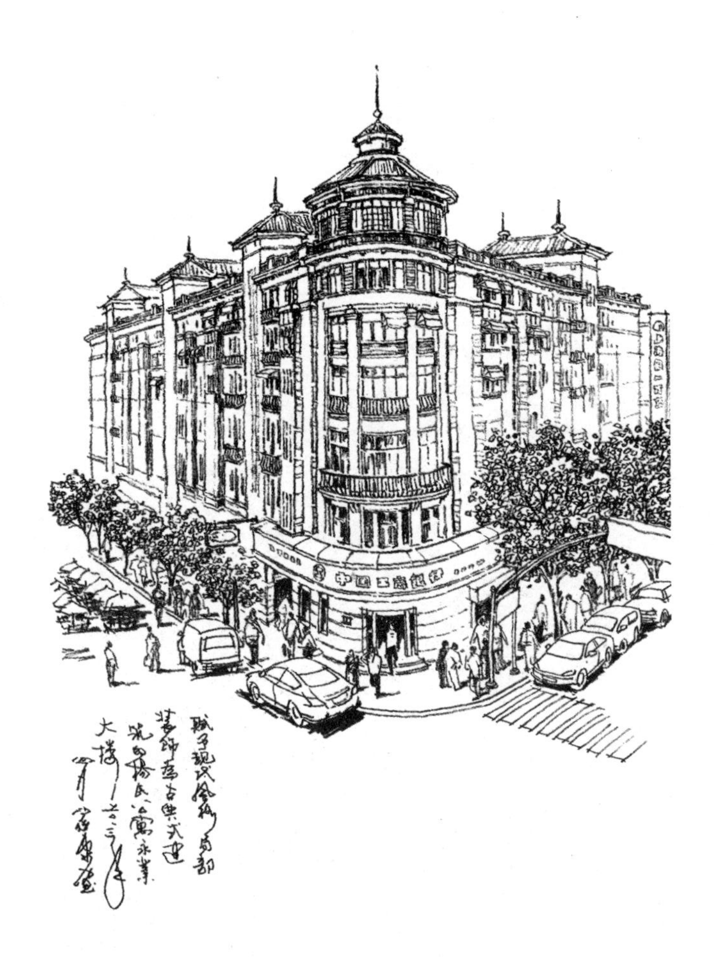

延庆路法国古典式花园住宅

（延庆路 130 号）

这栋巴洛克风格的花园住宅建于 1923 年，曾是一位英国籍犹太富商的住宅。

Classic French Garden House on Yanqing Road

This used to be the home of an affluent British Jewish merchant.

Address: No.130, Yanqing Road.

布哈德住宅

（永福路 52 号）

这栋漂亮的西班牙建筑建于 1932 年，曾是法国传教士布哈德的住宅。

Former Residence of Buchard

This elegant Spanish building was the house of French missionary Buchard.
Address: No.52, Yongfu Road.

永福路西班牙式花园住宅

（永福路 151 号）

这栋西班牙风格的花园住宅建于 1935 年，曾是德国驻上海总领事的官邸。

Spanish Garden House on Yongfu Road

This was an office building of the former German Consulate General in Shanghai.

Address: No.151, Yongfu Road.

孔祥熙旧宅

（永嘉路 383 号）

这座英国乡村式花园建筑是孔祥熙在上海的三处豪宅之一，始建于 1926 年，由华人著名建筑师范文照设计。

Former Residence of Kung Hsianghsi

This was one of Kung Hsianghsi's three mansions in Shanghai.

Address: No.383, Yongjia Road.

荣智勋住宅

（永嘉路 389 号）

这幢英国乡村式花园住宅竣工于 1936 年。窗间的几何图案墙饰系典型的装饰艺术风格。其原为比利时路易士洋行所有，后为荣智勋居住。

Former Residence of Rong Zhixun

This house with a British countryside-style garden used to belong to Lewis Bank of Belgium; it later became the residence of Rong Zhixun.

Address: No.389, Yongjia Road.

永嘉路花园住宅

（永嘉路 495 弄 1—9 号）

这幢欧式风格住宅建造于 20 世纪 30 年代，其外墙面用水泥砌出鱼鳞状细纹，非常独特。

Garden House on Yongjia Road

Its walls have unique scale-like veins.
Address: No.1-9, 495 Lane, Yongjia Road.

宋子文旧居

（永嘉路 501 号）

这幢德国城堡式花园住宅，建于1928年，是宋子文的故居之一。

Former Residence of Soong Tse-ven

This German garden castle was one of Soong Tse-ven's residences.

Address: No.501, Yongjia Road.

外国弄堂“雷米坊”

（永康路 109 弄）

雷米坊竣工于 1931 年。建筑配有独立花园和毗连式单体车库，底层多采用挑空设计。这里曾是各大洋行的外籍职员聚居地。

Remi Apartments, Foreign Lane

This used to attract staff from major foreign firms, hence the name “Foreign Lane”.
Address: 109 Lane, Yongkang Road.

岳阳路现代式花园住宅

（岳阳路 110 号）

这幢现代式花园住宅建造于 20 世纪 30 年代，为二层砖木结构，平面呈凹字形。

Modern House on Yueyang Road

This is a two-story concrete and wood structure.

Address: No.110, Yueyang Road.

宋子文旧宅

（岳阳路 145 号）

这幢荷兰式花园住宅建于 1928 年，是宋子文的别墅之一。当年宋子文在南京任国民政府财政部部长时，与大多数官员一样，把家眷安排在上海的租界里。

Former Residence of Soong Tse-ven

It was one of the villas owned by Soong Tse-ven.

Address: No. 145, Yueyang Road.

岳阳路现代式花园别墅

（岳阳路 170 弄 1 号）

这幢住宅建于 1936 年。其立面通过位于二楼的长形阳台、三楼的半凹凸阳台以及西侧的半圆形平台，创设出空间上的错落层次。

Modern Garden on Yueyang Road

The balcony of this building is quite unique.

Address: No.1, 170 Lane, Yueyang Road.

霖生医院旧址

（岳阳路 190 号）

这幢大型英式花园住宅建于 1920 年。民国时期名噪一时的“霖生医院”即创办于此。淞沪抗战爆发后，宋庆龄、何香凝等曾在此组织战地救护。

Former Linsheng Hospital

This hospital served refugees during the Anti-Japanese War (1937-1945).
Address: No.190, Yueyang Road.

陈楚湘住宅

（愚园路 395 弄涌泉坊 24 号）

出身于烟草世家的陈楚湘，1924 年继承父业，创办福和烟草公司，任总经理。这幢富有童趣的建筑，是他在 1936 年投资建造的。

Former Residence of Chen Chuxiang

Chen Chuxiang was a famous patriotic merchant.
Address: No.24, Yongquan Mill, 395 Lane, Yuyuan Road.

蒋光鼐旧居

（愚园路 1112 弄 4 号）

著名爱国抗日将领蒋光鼐旧居，建于 1928 年，属巴洛克建筑风格。

Former Residence of Jiang Guangnai

Jiang Guangnai was a famous general in the Anti-Japanese War (1937-1945).

Address: No.4, 1112 Lane, Yuyuan Road.

王伯群住宅

（愚园路 1136 弄 31 号）

这座意大利哥特式城堡建于 1934 年，曾是国民政府交通部长兼大夏大学董事长王伯群的住宅。

Former Residence of Wang Boqun

Wang Boqun was the Minister of Communications in the Republic of China Period (1912-1949).
Address: No.31, 1136 Lane, Yuyuan Road.

董竹君住宅

（愚园路 1294 号）

这栋建筑属英国城市别墅风格，建于 1930 年。原主人董竹君是上海锦江饭店创始人，她的一生非常传奇。

Former Residence of Dong Zhujun

Founder of Shanghai Jin Jiang Hotel, Dong Zhujun led a legendary life.

Address: No.1294, Yuyuan Road.

新华村

（愚园路 1320 弄）

当年的新华村由五幢英国城市别墅建筑组成。外墙面典雅的河卵石，直通二楼居室的楼外阶梯，均彰显出建筑的与众不同。该楼建于 1925 年，本系外侨私产，后几易其主。

Xinhua Village

These villas, which belonged to an oversea Chinese merchant, were resold several times and are now government buildings.

Address: 1320 Lane, Yuyuan Road.

西本愿寺

（乍浦路 455 号）

这原是 20 世纪上半叶日本佛教庙宇西本愿寺在上海的别院，呈现印度佛教建筑特征。

West Honganji Monastery

It is a branch of Japanese Nishi Honganji in Shanghai, showing a style of the Buddhist architecture in India.

Address: No.455, Zhapu Road.

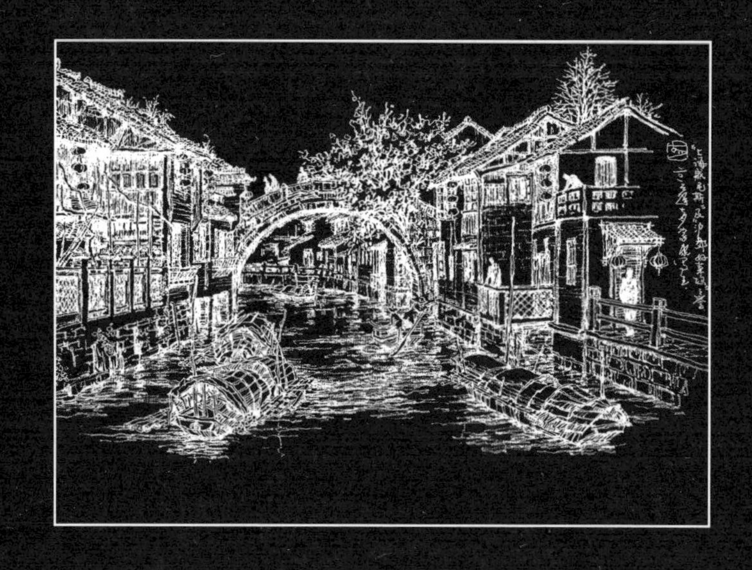

闵行区、青浦区、崇明县等

Minghang District, Qingpu District, Chongming County, etc.

法华塔

（嘉定区嘉定镇南大街 349 号）

法华塔，又名金沙塔，位于嘉定州桥老街，始建于宋代开禧年间，历代均有不同程度的修缮。1996 年清理发掘该塔地宫，发现众多珍贵文物。

Fahua Tower

Fahua Tower is over a thousand years old; many precious relics have been excavated from here.

Address: No.349, South street Jiading Town, Jiading District.

七宝古镇

（闵行区七宝镇）

七宝镇位于上海市西南部，是一座有着千年历史的江南水乡古镇。

Qibao Anccient Town

Located in northwestern Shanghai, Qibao has a style typical of architecture south of the Yangtze River, enjoying a history of over a thousand years.

Address: Qibao Town, Minghang District.

南张天主堂

（闵行区莘庄镇明星村）

此天主堂建于清光绪二年（1876），本名为“若瑟善终立保堂”，因地处莘庄乡明星村南张地区（今莘庄镇秀文路），故俗称“南张天主堂”。

Nanzhang Cathedral

This former catholic church is now a home for the aged.

Address: Mingxing Village, Xinzhuang Township, Minghang District.

朱家角古镇

（青浦区朱家角镇）

朱家角古镇历史悠久，有“上海威尼斯”之称。临近淀山湖风景区，与大观园风景区隔湖相望。

Zhujiajiao Ancient Town

Zhujiajiao,enjoying a long history, is also called “Shanghai Venice”.

Address: Zhujiajiao Town, Qingpu District.

泰来桥天主堂

（青浦区环城镇城南村）

泰来桥天主堂始建于清同治四年（1865），1927年改建为哥特式大堂。20世纪50年代后停止活动，1988年修复开放。教徒多为当地渔民。

The Catholic Church by Tailai Bridge

The believers here mostly consist of local fishers.

Address: Chengnan Village, Huancheng Town, Qingpu District.

练塘灵恩堂

（青浦区练塘镇下塘街 37 弄 5 号）

练塘灵恩堂始建于清光绪二十九年（1903）。约于 20 世纪 50 年代，该堂停止宗教活动，建筑挪作他用。1986 年恢复活动，2000 年扩建。

Ling'en Church in Liantang

The church was built in 1903 and expanded in 2000.

Address: No.5, 37 Lane, Xiatang Street, Liantang Town, Qingpu District.

蔡家湾天主堂

（青浦区徐泾镇高经村 185 号）

蔡家湾天主堂初建于清雍正十二年（1734），时名“慈母堂”。清道光三十年（1850）附设孤儿院，传授孤儿手工技艺。该堂还曾创办缝纫、木工、制鞋、印刷等工场。

CAI Home Bay Church

Built in 1734, the church was originally named “mother’s hall”.

Address: No.185, Gaojing Village, Xujing Town, Qingpu District.

佘山天主教堂

（松江区佘山镇）

这幢褚红色天主教堂位于佘山山顶，被誉为“东亚第二大教堂”，是国内天主教最主要的朝圣地。清同治十年（1871）由法国传教士始建，1925年翻造扩建，1935年落成。

Sheshan Cathedral

It is the most important pilgrimage site of catholic church in China.

Address: Sheshan Town, Songjiang District.

黄家花园

（崇明县城桥镇）

崇明岛上的黄家花园是当地富商黄稚卿于 1927 年所建。该建筑属中西合璧式花园别墅。

Garden of the Huang Family

This garden was built by Huang Zhiqing, an affluent Chongming merchant.

Address: Chengqiao Town, Chongming County.